Zelia Missagia
André Luis Christoforo
Francisco Antonio Rocco Lahr

Composites Reinforced by Sawdust, Cement and Magnesium Silicate

AF153195

Zelia Missagia
André Luis Christoforo
Francisco Antonio Rocco Lahr

Composites Reinforced by Sawdust, Cement and Magnesium Silicate

Engineering Materials

LAP LAMBERT Academic Publishing

Impressum / Imprint

Bibliografische Information der Deutschen Nationalbibliothek: Die Deutsche Nationalbibliothek verzeichnet diese Publikation in der Deutschen Nationalbibliografie; detaillierte bibliografische Daten sind im Internet über http://dnb.d-nb.de abrufbar.
Alle in diesem Buch genannten Marken und Produktnamen unterliegen warenzeichen-, marken- oder patentrechtlichem Schutz bzw. sind Warenzeichen oder eingetragene Warenzeichen der jeweiligen Inhaber. Die Wiedergabe von Marken, Produktnamen, Gebrauchsnamen, Handelsnamen, Warenbezeichnungen u.s.w. in diesem Werk berechtigt auch ohne besondere Kennzeichnung nicht zu der Annahme, dass solche Namen im Sinne der Warenzeichen- und Markenschutzgesetzgebung als frei zu betrachten wären und daher von jedermann benutzt werden dürften.

Bibliographic information published by the Deutsche Nationalbibliothek: The Deutsche Nationalbibliothek lists this publication in the Deutsche Nationalbibliografie; detailed bibliographic data are available in the Internet at http://dnb.d-nb.de.
Any brand names and product names mentioned in this book are subject to trademark, brand or patent protection and are trademarks or registered trademarks of their respective holders. The use of brand names, product names, common names, trade names, product descriptions etc. even without a particular marking in this work is in no way to be construed to mean that such names may be regarded as unrestricted in respect of trademark and brand protection legislation and could thus be used by anyone.

Coverbild / Cover image: www.ingimage.com

Verlag / Publisher:
LAP LAMBERT Academic Publishing
ist ein Imprint der / is a trademark of
OmniScriptum GmbH & Co. KG
Heinrich-Böcking-Str. 6-8, 66121 Saarbrücken, Deutschland / Germany
Email: info@lap-publishing.com

Herstellung: siehe letzte Seite /
Printed at: see last page
ISBN: 978-3-659-71399-6

Zugl. / Approved by: São Joao del Rei, Federal University of São Joao del Rei, Dissertation in Mechanichal Engineering (PPMEC-UFSJ), 2013.

Composites Reinforced by Sawdust, Cement and Magnesium Silicate

Zélia Maria Velloso Missagia
André Luis Christoforo
Francisco Antonio Rocco Lahr

i

"If you want to change the world, first, trying to promote your personal development and create innovations in yourself"

Dalai Lama

ABSTRACT

This study aims to investigate the physical properties (density, apparent density, water absorption) and mechanical properties (elasticity modulus and compressive strength) of composites in particulate reinforced epoxy sawdust, Portland cement and magnesium silicate. An initial experimental design was developed involving only the epoxy resin and sawdust. The factors investigated in this design were: wood species (*Eucalyptus grandis-Eucalyptus* and *Tabebuia serratifolia* - Ipe), particle size range (4-10and 50-80US-Tyler) and mass fraction of sawdust on the resin (30%, 50%) providing eight different experimental conditions. Subsequently, the treatment was chosen which gave the highest values for the elasticity modulus in compression. This chosen condition, four other conditions were developed, incorporating 10 to 20%inmass fraction of particles of magnesium silicate, and Portland cement (PC-V) on the resin. Compounds with 30% *Eucalyptus* wood sawdust and US-range 50-80Tylersieveshowed the best values for the elastic modulus in compression, being used as a reference condition for the inclusion of cement and talc particles. The use of cement in the compounds gave significant increases in the compressive elasticity modulus, apparent density and with reduced water absorption, which does not occur with the incorporation of magnesium silicate particles, presenting be significant only in absorption, being the highest values resulting from materials manufactured by adding 20% of the reinforcement particles. It follows that the material can be classified as compressive strength as allow-strength concreteor concrete non structural low weight and, adding a second phase particles contribute to an increase the issues of the modulus (as desired regardless of the increase) with the possibility of losing other properties.

KEYWORDS: Wood, compression, strength, design of experiments (DOE).

LIST OF FIGURES

LIST OF TABLES

LIST OF ABBREVIATIONS

BS: Regulatory British Standards;

CITEC: Innovation and Technology Center in Composites;

FRP: Reinforced Polymers with Fibers;

MEV: Scanning electron microscopy;

Minitab: Software for statistical analysis;

ABNT NBR: Brazilian Association of Technical Standards;

PP: polypropylene;

P-value: Statistics used to synthesize the result of a hypothesis testing;

PVC: Polyvinyl chloride (plastic);

UFSJ: São João del-Rei Federal University;

US-Tyler: Mesh size for selection;

WPC: Wood Plastics Composites;

LIST OF SYMBOLS

k: Kilo (one thousand units of magnitude);

M: Mega (one million units of magnitude);

m_1: Initial mass;

m_{12}: Sample mass at 12% of moisture;

MOE: Elasticity modulus;

MOR: Strength modulus;

m_s: Dry mass of sample;

η: Porosity;

v_{12}: Sample volume at 12% of moisture;

v_{sat}: Volume of saturated sample;

v_t: Total volume of sample;

v_v: Voids in the sample.

SUMMARY

Chapter 1

INTRODUCTION

The waste from the wooden processing has emerged as aggravating contribution to the generation of environmental impacts, encouraging researches that seek solutions to this problematic of global dimensions.

Among them, stand out the wood as sanding dust and sawdust, to be materials of low density, requiring more space for storage and are highly explosive (YAMAJI and BONDUELLE, 2004).

There is currently a major expansion in the use of waste to provide a better use of wood, enabling a lossless quality of the final product processing and at the same time, alerting the population so that there is an appropriate destination for the same (SALVASTANO, 1998; MIOTTO and DIAS, 2006; PASSEROTTI *et al*, 2008).

An example of materials based on wood are called wood-plastic composite (wood plastic composites - WPC), prepared from the wood dust composition with some type of plastic resin (KOENIG and SYPKENS, 2002), consisting of an innovative product by offer aesthetic finish and beauty to the environment, benefit consumption and commitment to the preservation of the planet. These materials replace conventional wood with significant improvements being environmentally friendly by having wooden waste as the main raw material (PASSEROTTI *et al*., 2008).

The WPC have been developed for various purposes (STARK, 2001), generally characterized in flexion (MISSAGIA *et al*., 2011), presenting some advantages over solid wood such as not crack and not bend, still requiring little or no maintenance over its lifetime (BRANDT and FRIDLEY, 2003).

These and other researches aim to spread the use of wooden waste in the development of new materials, whose physical and mechanical properties, depending on the factors and stipulated experimental levels, can reach or even exceed those of solid wood (ENGLISH, 2002) showing up, in addition to other applications, such as alternative solutions in the form of repair or reinforcement in structural timber, furniture etc. (YAMAJI and BONDUELLE, 2004).

1.1. Objectives

This study aimed the development and the physical and mechanical characterization (compression) of composite materials in reinforced matrix epoxies with sawdust, cement and magnesium silicate, allowing investigate their potential in form of repair or reinforcement either in parts or timber structures.

1.2. Justification

The motivation for the development of this work was the possibility of utilization of wooden solid waste, usually discarded in nature (YAMAJI and BONDUELLE, 2004), in development of wood-plastic composite materials, contributing to increase the range of knowledge about their physical and mechanical properties, as well as for spreading the use of sawdust in the development of new materials.

Chapter 2

LITERATURE REVIEW

In the literature review sought to present researches on the development of composite materials made from sawdust and resin, types of treatments used to improve the interface conditions of these materials and the usual forms of their employment being used for both, the databases: Web of Science, Engineering Village, Portal Capes, Scielo, Science Direct, among others.

2.1. Composites manufactured from wooden sawdust

The employment of natural fibers as reinforcement in polymer composites is justified basically by the possibility of obtaining good physical and mechanical properties, reduction of the final cost (when compared with materials made with synthetic fibers) and are biodegradable, by helping to mitigate the environmental impacts (CARVALHO, 2003; MCHENRY and STACHURSKI, 2003; PANTHAOULAKKAL *et al*, 2005;. PASSEROTTI et al, 2008).

The composite comprising of wooden dust, phenolic resins became known as Bakelite, a thermosetting resin which was the first plastic synthesized industrially. This revolutionized product design, especially in manufacturing home appliances in the 20 to 50 decades, whereupon the new plastics have replaced (Teixeira, 2005).

In industry of processing materials, developed studies revealed that the use of natural fillers in composite materials represent an increase of 60% in applications within the

Literature Review

automotive industry, especially in materials that use PVC as a matrix of natural fibers (HRISTOV *et al.* , 2004; TORRES and CUBILLAS, 2005; CRESPO, 2006).

Faced with this evidence, it is noted that social concern and existing policy with environmental protection have been reflected in the high protection of forest resources. Thus, the combination of natural and synthetic fillers polymers, in order to manufacture inexpensive products that replace traditional wooden materials offer a very wide range of applications (PANTHAPULAKKAL *et al.*, 2005).

Historically, the use of wooden particles as reinforcement for thermoplastics has been reported by several authors (BLEDZIK *et al.*, 2005).

English*et al.* (1997) fabricated a composite material from waste wood flour of Pine wood and demolition wood with synthetic resin (polypropylene) and other reinforcements (talc, glass fiber, calcium carbonate) using mass fractions of 20 and 40% of wood flour on the resin, with the aim of reducing the specific weight of the transport packaging. The results showed that the glass fiber associated to fraction of 40% residue gave the best values of tensile strength (39 MPa) and elasticity modulus (3.80 GPa) and bending strength (36 MPa), even in the best consisting condition for the specific weight.

Clemons (2002) discusses the use of wooden powder or fibers of any kind in a proportion of 2 to 50% in mass fraction on thermoplastic resins (polyethylene, and polystyrene, polypropylene and polystyrene) allow an increase in the bending stiffness properties (3.22 GPa), so that they can be used in construction profiles with windows and decks, it also show a significant improvement in the permeability of the material.

Correa *et al.* (2003) evaluated the mechanical properties of strength and stiffness to bending in high-impact polystyrene composite (HIPS) with the inclusion of three waste types of Pinewood with and without treatment with maleic anhydride (PP-MAH). Authors found positive action of PP-MAH compatibiliser by increasing of stiffness modulus and tensile strength (3.0 GPa) made compatible mixtures of polypropylene wooden flour, independent of the particle size range of wood. Results of tensile tests demonstrated the

Literature Review

positive action of PP-MAH compatibiliser by increasing the stiffness modulus and tensile

strength (29.6 MPa) of made compatiblemixtures of polypropylene and wooden flour

Khouylou (2006) fabricated and compared the physical and mechanical properties of

composite materials with cementations binders (cement with sand and water) and sawdust

undefined species impregnated with unsaturated polyester resin containing styrene and/or

methyl methacrylate (MMA) followed by exposure to gamma radiation from Cobalt-60.

These results showed that the compressive strength and bending modulus were compared to

those of high-quality concrete, being its porosity significantly reduced.

Salemane *et al.* (2006) in their study demonstrated that when a sawing (WP) is used

as a reinforcing agent in addition to plastic materials (PP), this tends to increase the

stiffness, does not providing improvements in strength. Also stated that the very end of

sawdust particles are difficult to disperse because agglomerates are formed, on this account

are formed agglomerated that will behave as large particles and that the possible chemical

treatment with maleic anhydride (MAPP) for improving the dispersion and interface

conditions may harm particles, particularly fine and can strongly influence the final

properties of composite, as discussed in the work Tang (1997).

Ashori and Nourbakhsh (2009) investigated the use of recycled wooden fibers and

plastics in the production of WPCs (Wood Plastic Composite). The polymers used were the

recycled high density polyethylene and polypropylene. The lignocellulose material for this

study consisted of old newsprint fibers. The manufactured panels were hot pressed and the

evaluation of physical properties of density and water absorption showed that the use of

polypropylene as the coupling agent improved the interface between constituent elements. It

was observed that composites made from high density polypropylene promoted moderate

superior mechanical properties (Bending modulus> 2.0 GPa and Elasticity modulus about

670 MPa), if compared with composites made from pure polypropylene.

Hisham *et al.* (2011) developed composite materials in epoxy resin reinforced by

wooden waste products obtained in the wooden industry originating from various types of

wood in three different sizes, being characterized in tension. The variance analysis to be

Literature Review

charged significant the size of fibers in the elasticity modulus and tensile strength, presenting higher values than compounds manufactured with longest fibers.

2.1.1. Treatments for improving the interface conditions

The fiber-matrix adhesion is one of the most important points to consider on the mechanical properties of a composite material, because the incompatibility between phases can result in inferior mechanical properties.

Actually, the methods used to improve fiber matrix adhesion include: entanglement of molecular chain, good mechanical contact, correlation of surface stresses and formations of chemical bonds through the use of chemical coupling agents (BLEDZKI and GASSAN, 1999; MOON *et al.*, 2005).

Joseph *et al.* (1999) investigated the effect of using ultraviolet (degradability) of physical properties (moisture content) and mechanical (modulus and tensile strength) of composites with polypropylene matrix reinforced by sisal fibers. Among others, these authors concluded that the use of gamma rays has provided an effective alternative method to match the two constituents.

Dibenedetto (2001) discusses the greater difficulty in making composites such as natural fibers is the fiber/matrix adhesion, which can be improved with the use of coupling agents such as amorphous silica, which modify the fiber surface by reducing the interfacial energy.

Yuan et al. (2004) fabricated composites with polypropylene matrix reinforced by Pinewood fibers, which after mixed were hot-pressed, which investigated the effect of surface treatment of fibers by plasma. Results obtained of microscopic analysis indicated that the plasma treatment is able to improve the compatibility of fibers with the polypropylene.

Kamel *et al.* (2007) investigated the physical and mechanical properties of composites made with sawdust and Pinewood of low density polyethylene, using maleic anhydride as a treatment of particles. Results of the bending and tensile tests showed the effectiveness of the treatment, still presenting materials with maleic anhydride lower values of water absorption.

Nassar (2007) evaluated the physical and mechanical properties of composites made with fibers of Pine sawdust, rice husks, and epoxy resin using chemical treatment with ethylene vinyl acetate. Results obtained by scanning electron microscopy (SEM) and by physical tests showed the efficiency of chemical treatment used.

Ahmad *et al.* (2008) developed a composite with sawdust of Acacia wood, unsaturated polyester resin (UPR) and recycled PET with alkaline polymerization treatment and glycolysis. Strength and stiffness in flexion and strength and water absorption were evaluated. Results showed that tensile and bending modulus increased (2.4 GPa), but the tensile and bending strengths decreased (from 28.9 MPa to 22.4 MPa). The size of sawdust particle performed a significant role in these mechanical properties. The authors concluded that the treatment used caused a better adhesion between the matrix and the sawdust, while reducing the water absorption of the composite.

Ku *et al.* (2009) developed composite materials with phenolic resin and three particle sizes of sawdust, being added Garamite and propylene glycol to improve the interface conditions. Among others, the authors concluded that the chemical treatment provided an increase in stiffness properties of manufactured compounds.

Raman *et al.* (2008) and Raman *et al.* (2010) stated that the mechanical properties of natural fibers in reinforced polymer composites can be significantly improved by pretreatment of fibers with sodium iodate and after treatment with urea urotropin.

Ku *et al.* (2011) evaluated the physical and mechanical properties of epoxy matrix composite materials reinforced with sawdust, being used two different heat treatments to improve the interface condition, one by oven (60°C) and the other by microwave. Among others, the authors concluded that the chemical treatment provided an increase in stiffness properties (approximately 2.0 GPa) of the manufactured compounds. From this point, one reached the conclusion that the fracture strength increased with the load increase of particles. These properties are vital for civil engineering applications, because the civil structures need composites with high stiffness and fracture toughness.

Alternatively, waste wood have been used in the manufacture of composites intended for various technological applications: in the automotive industry, construction (repair and reinforcement structures), in toys, outdoor use materials (furniture) and also in packaging in general (KURUVILLA *et al.*, 1999; AL-QUERISH *et al.*, 1999).

2.1.2. Applications of materials fabricated with resin and sawdust

AIJWE *et al.* (1998) produced epoxy resin composite materials, rice straw and sawdust for employment as for residential roofing tile, being the mechanical and physical properties compared to those from ceramic tiles manufactured on an industrial scale. The results indicate that the mechanical properties of the manufactured materials were consistent with those of ceramic tiles, and that the addition of the straw afforded significant reduction in water absorption.

Dagher *et al.* (2002) studied comparatively three reinforcement technologies in timber beams for six years using laminates reinforced composites with fiberglass, carbon fibers and natural fibers, concluding that, in accordance with the design and the proper use, the three types of reinforcement that were viable of mechanical and economic point of view.

The use of recycled plastic for the manufacture of WPC's (Wood Plastic Composites) has been studied by a large number of authors with applications in the form of external and internal floors, vases, baskets, benches, tables, drums, etc. (AVILA and DUARET, 2003 ; JAYARAMAN and BHATTACHARYYA, 2004).

Regarding the civil construction, in structural designs should be evaluated the durability characteristics of the materials, in the same way as their costs, so that resist without deterioration for many years. Interventions are not only to restore the load capacity of the structure (repair) or to increase the load capacity of the (reinforcement), but also to minimize the action of external aggressors to the material used in the manufacture of structural elements (GARCEZ *et al.*, 2004).

Literature Review

Researches and studies in reinforcement of timber structures with natural fibers are in the initialization phase. The advantages presented (abundance, biodegradability, low or no cost) with respect to synthetics, the latter being still the most used (MIOTTO *et al.*, 2006), it is believed that the development of new works will highlight the employment potential of natural fibers as reinforcement structures (SANCHES, 2001).

The biological attack is a major cause of degradation in timber structures, resulting in mass loss and hence of strength (MIOTTO *et al.*, 2006).

Design and/or construction errors, degradation and aging of materials, changes in dimensioning codes (dispositions more severe or occurrence of accidents) are the main factors that has motivated the development of researches to this problem, especially the use of composites, due to their good relationship between strength and density, easy to perform (the architectural configuration and aesthetic issues are little affected) and are generally immune to corrosion (RANGEL, 2010).

Khoo (2008) developed composite materials with sawdust, phenolic resin and rubber to be used in exposed timber columns to the elements, being investigated the properties: elasticity modulus and compressive strength. Two medium-sized were used in (2,515 mm) and (4,826 mm) and of rubber and two mass fractions of waste on the resin (10 to 40%). The results of the mechanical properties indicated the feasibility of using the composites manufactured in timber to repair columns.

Vijay (2011) developed composite materials with urea-formaldehyde resin and sawdust for recovery columns (subject to weather) of timber bridges of South Branch Valley Railroad. The repair could be applied without being interrupted the regular traffic on the bridge. After fixed the compounds in damaged columns, these were subsequently wrapped in a laminate fiberglass composite. After restoring NDT (not destroyed tests) evaluation were performed. These tests have shown improving in strength and stiffness.

Sales *et al.* (2011) developed composite materials made with Pine wood waste, "sludge" resulting from water treatment company in São Carlos City and cement paste (cement, water and sand). The results of compressive strength, water absorption and density

Literature Review

were equal to 1.2 MPa, 8.8% and 1,847 kg/m^3, classifying them as qualified for structural repairs.

The production of particleboard homogeneous with wood particles of Amazon region of low and medium densities (*Erisma uncinatum*, *Nectranda lanceolata*, *Erisma sp*) were studied by Silva and Lahr (2007). In evaluations performed, in accordance with the NBR 14810:2002b of the Brazilian Association of Technical Standards (ABNT), the plates of *Nectranda lanceolata* wood particles showed the highest strength values in bending, which are higher than the limit established by this standard.

Dias *et al.* (2008) studied mechanical properties of particleboard wood panels made of polyurethane resin on mammon base. The results obtained for the elasticity modulus in bending did not reach the minimum value of 18 MPa, possibly explained by the poor distribution of the adhesive during the panel forming process.

Akgüla and Çamlibelb (2008) studied the strength and stiffness of particles produced with *Rhododendron* wood panels, and the moisture content of 14% and particles together by adhesive based on urea-formaldehyde. The results indicate the investigated conditions the use of wood in the manufactured *Rhododendron* panels.

Also, the use of sawdust *Pinus elliottii* and polyurethane resin mammon-based have great employment in manufacturing of fibersheets and particleboard (PAES *et al.*, 2011).

The feasibility of production of particleboard in rubber tree clones RRIM 2020 with four years of age were studied by Saffian *et al.* (2011), being evaluated the elasticity modulus (MOE) and of rupture modulus(MOR). The results indicated that be possible to manufacture panels with rubber clones evaluated.

Hisham (2011) conducted studies with particleboard of wood with industry waste (various species), and three particle size ranges and two mass fractions of epoxy resin (10 and 20%) on fibers, and pressed (3 MPa) cold. The results of physical and mechanical properties were satisfactory, both being equivalent to those obtained by other authors.

Paes*et al.* (2011) studied the effect of the combination of pressure and temperature on particleboard with *Pinus elliottii* wood waste and polyurethane resin derived from the mammon in properties: density, swelling and water absorption; elasticity modulus and rupture to bending, screw pullout and internal connection, concluding that the combinations 3.0 MPa and 90°C and 3.5 MPa and 60°C showed the best results, proving to be the temperature of pressing the most significant variable for quality the prepared plates.

2.2. Conclusions of literature review

The use of natural fibers as reinforcement in composite materials has been the focus of several studies, motivated in an attempt to replace the commonly manufactured compounds with synthetic reinforcements, because presented the fibers, as a good relation, concerning mechanical strength and weight, combined to sustainability issues.

Strength and bending stiffness were the most exploited mechanical properties in research involving the composite resin and sawdust, being few works developed, in which were investigated the strength and stiffness in uniaxial compression, which are a major objectives of this research.

The treatment for improving the adhesion between resin and wood fibers was the focus of several studies, being used physical and chemical treatment methods, with favorable results in many of the papers presented, however, not being used in this work, no method for the treatment surface of sawdust particles.

The joint use of particles and wood fibers as a reinforcing phase in composite resin (matrix) has been exploited by some researchers, comprising the addition of particles (cement and magnesium silicate) in a factor to be investigated one valuated physical and mechanical properties.

The versatility of use of composite materials made with resin and sawdust is evidenced by the listed works, and may be used in the form of tile, repair or reinforcement in structures and furniture in general, floors, panels (various applications), automotive vehicles component parts among others, further encouraging the development of new works that contribute for fastening the use of wood waste in the development of new materials.

Chapter 3

MATERIAL AND METHODS

In this chapter are described the methods of manufacture of the materials, the physical and mechanical tests used (supported by the use of normative documents) and the statistical approach employed in the analysis of the results.

For the evaluation of physical and mechanical properties of composite materials made from epoxy resin, sawdust, cement and magnesium silicate two separate studies were performed. At first, there was an experimental planning using only the epoxy resin and sawdust, and investigated the condition relating to the greater elasticity modulus were incorporated magnesium silicate particles (talc) and Portland cement.

The factors stipulated in the study of compounds manufactured with resin and sawdust were: wood species (*Eucalyptus grandis* - *Eucalyptus*; *Tabebuia serratifolia* - Ipê), particle size range (4-10; 50-80 US-Tyler) and mass fraction of sawdust on resin (30%; 50%), leading to a complete factorial planning of the type 2^3, providing eight experimental conditions (EC) distincts, explained in Table 3.1.

TABLE 3.1: Conditions of experimental planning of materials made with resin and sawdust.

CE	Type of sawdust	Particle size range (US-Tyler)	Sawdust fraction (%)
C1	*Eucaliptus*	4-10	30
C2	*Eucaliptus*	4-10	50
C3	*Eucaliptus*	50-80	30
C4	*Eucaliptus*	50-80	50
C5	Ipê	4-10	30
C6	Ipê	4-10	50
C7	Ipê	50-80	30
C8	Ipê	50-80	50

The response-variables investigated in the materials of the eight experimental conditions of Table 3.1 were: apparent density (ρ_{ap}), water absorption in 24 hours, the elasticity modulus in compression (MOE) and strength modulus in compression (MOR).

The variance analysis (ANOVA) was used to investigate the influence of individual factors (wood species; particle size range; mass fraction sawdust), as well as the interaction of both the physical and mechanical properties of interest.

According Werkema and Aguiar (1996), a factor or the interaction between two or more factors is considered significant when the P-value obtained from the ANOVA results in a number less than 0.05 (5%). The authors also argued that the analysis of the interaction between the factors involved is more important than the individual, making it necessary in these cases only investigation of the effects of interaction. Accused significant a factor or the interaction between them in the responses investigated by ANOVA were employed posteriorly the comparison test of Tukey Averages.

Finding among the eight treatments, the condition of greater value for the elastic modulus in compression, to this are added separately Portland cement (PC-V AIRI) and

Material and Methods

magnesium silicate mass into two fractions on the resin, with 10% and 20%. The ANOVA was employed to assess the influence of inclusion of two fractions of cement particles in the compound of higher MOE (reference), and later used the Tukey Test. The same approach was used for statistical analysis of the influence of the addition of the magnesium silicate particles in the compound of higher elasticity modulus. The response-variables investigated of these materials are the same of resin-sawing materials, except for compressive strength module.

3.1. Constituent materials

The *Eucalyptus* wood and Ipê waste were provided by Agostini Sawmill and Lumber Industry, timber industry companies in the municipality of São João del Rei/MG.

The sawdust obtained was sifted for five minutes following the recommendations of the American ASTM D1921:2012, with the aid of a shaker of vibrating sieves (Figure 3.1) available in Materials Laboratory of Department of UFSJ Mechanical Engineering, in order to classify waste in 4-10 and 50-80 US-Tyler particle size ranges. Figures 3.2 and 3.3 respectively illustrate the *Eucalyptus* and Ipê the saw dusts for both investigated particle size ranges.

FIGURE 3.1: Shaker of vibrating sieves.

(a)   (b)

FIGURE 3.2: Sawdust *Eucalyptus* in particle size ranges 4-10 (a) and 50-80 US-Tyler (b).

(a)   (b)

FIGURE 3.3: Ipê sawdust in particle size ranges 4-10 (a) and 50-80 US-Tyler (b).

Concluded the classification of waste, the same were conducted in an oven to correct the moisture content by 12%, following the assumptions and guidelines of the Brazilian Standard NBR 7190:1997 (Design of timber structures).

Magnesium silicate (Figure 3.4-a) and Portland cement CP-V AIRI (Figure 3.4-b) used in the manufacture of composites were those of VERBAZZA and CAUE trademarks, respectively provided by trade companies of construction materials of São João del Rei City/MG.

Material and Methods

FIGURE 3.4: Magnesium silicate (a) and Portland cement CP-V (b).

The epoxy resin (Figure 3.5-a), Araldite-M of Hunstman® trademark and the RenShape HY 956 hardener (Figure 3.5-b) were provided by the World of Resin company, of Belo Horizonte /MG, using the proportions of 5 parts resin to 1 of catalyst, as specified by the manufacturer.

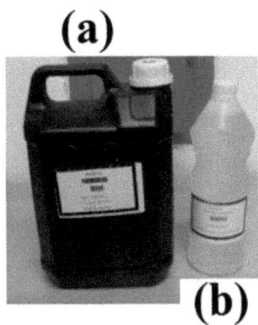

FIGURE 3.5: Resin (a) and catalyst (b).

3.2. Production of materials

Mixtures of materials resin-sawdust (Figure 3.6), the resin-sawdust-cement and resin-sawdust-talc were performed manually, and the particles proportions of 30% to 50% by mass fraction on the resin were well defined after conducting the initial tests (mixtures between phases). In these mixtures were adopted proportions in mass, instead in volume, by

Material and Methods

the great variability in the wooden density, potentially increased because it is waste. This practice is common in work involving the composition of wooden fibers and resin, however, differing slightly from the panels and particle plates, in which the adhesive fraction is defined on the mass of wooden fibers.

FIGURE 3.6: Mixture manual between Ipê sawdust and epoxy resin.

3.3. Production of specimens

Because of the mechanical properties investigated be obtained from the uniaxial compression tests and the physical not depend on the shape of specimens for both trials, it was used samples of cylindrical geometries, with 40 mm length (h) and 20 mm in diameter (d) as illustrated in Figure 3.7, keeping the relation h = 2 d established by the American Standard ASTM D695-10: 2010.

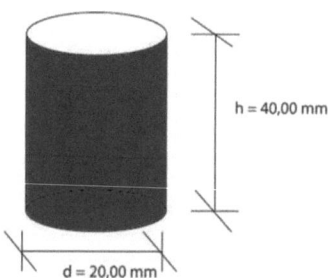

FIGURE 3.7: Geometry and dimensions of specimens.

Material and Methods

After mixed phases (Figure 3.8), the materials resulting from each experimental condition were inserted manually and gradually in PVC tubes (Figure 3.8-a) with 20 mm internal diameter and 60 mm in length, properly seated on a wooden support and compressed with the aid of a steel turning cylinder (Figure 3.8-b).

FIGURE 3.8: Formation of specimens.

After seven days (curing time) of the insertion of compounds in the PVC molds, the same were removed with the aid of an electric saw and subsequently taken to be finished (Figure 3.9).

Material and Methods

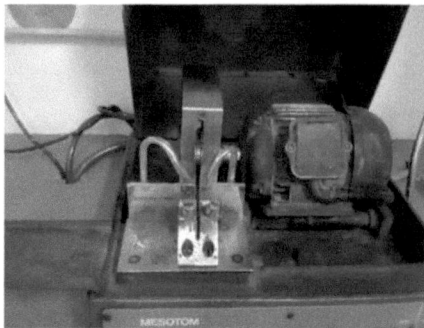

FIGURE 3.9: Facing of specimens.

The Figure 3.10 illustrates specimens made of the eight experimental conditions (Table 3.1) relating to compound resin-sawdust.

FIGURE 3.10: Compounds made of resin and sawdust.

Material and Methods

Figure 3.11 explains the sequence of steps involved in the manufacture of specimens.

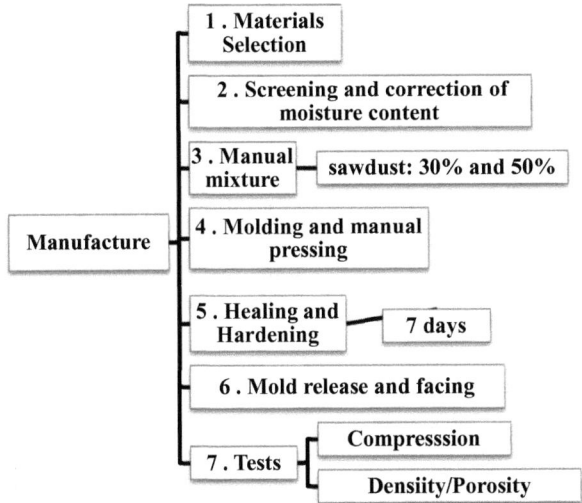

FIGURE 3.11: Flow chart of the steps involved in the manufacture of specimens.

3.4. Trials

For the achievement of physical and mechanical tests, by experimental condition (EC) investigated were manufactured 14 specimens, 10 for the compression tests and the remaining four for tests of volumetric density and absorption in 24 hours. For resin-sawdust compounds were manufactured 112 specimens, 28 for resin-sawdust-talc compounds and another 28 for composite resin-sawing cement, totaling 168 specimens, and 168 trials between physical and mechanics tests. The specimens per condition were made five to five on different days for mechanical tests and pairs to the physical tests (two replicates per experimental condition), enabling verify the homogeneity obtained in the process of preparation of materials.

3.4.1. Mechanical tests

The mechanical tests (Figure 3.12) consisted in obtaining the elasticity modulus (MOE) and strength modulus to compression (MOR). Therefore, were used the assumptions and calculation procedures described by the American ASTM D695-10:2010. Composite resin-sawdust materials were tested on the premises of Structural Engineering Department of Minas Gerais Federal University (UFMG), and the other being performed in Laboratory of Materials of Mechanical Engineering Department of São João del Rei Federal University (UFSJ).

FIGURE 3:12: Test of uniaxial compression.

The MOE and MOR of the compounds were obtained with the use of Equations 3.1 and 3.2 respectively, where m is the slope of the line set in the linear section of the diagram tension x deformation, F_{max} the maximum applied force and S the cross-sectional area of the specimen.

$$MOE = m \qquad (3.1)$$

$$MOR = \frac{F_{máx}}{S} \qquad (3.2)$$

3.4.2. Physical tests

As stated before, the physical tests consisted in obtaining the volumetric density (ρ_v) and water absorption after 24 hours (A_m) obtainable according to the methodological procedures in the Standards EN 323:1993 and EN 317:1993, respectively. To obtain the physical properties were utilized a digital caliper, an analytical balance accurate to 0.01g in a vacuum system (Figure 3.13).

FIGURE 3.13: Submersion of specimens.

The specimens after immersion in water stayed in vacuum pump at a pressure of 0.5 bars.

The volumetric density (Equation 3.3) is defined as the ratio between the mass ($m_{12\%}$) and the sample volume ($V_{12\%}$) for the 12% moisture content.

$$\rho_v = \frac{m_{12\%}}{v_{12\%}} \qquad (3.3)$$

Getting absorption in 24 hours (Am) was calculated according to equation 3.4, with the masses m1 and m2 respectively, measured before and after the immersion in water of the samples

$$A_m(\%) = \frac{m_2 - m_1}{m_1} \cdot 100 \qquad (3.4)$$

Chapter 4

RESULTS AND DISCUSSION

Table 4.1 presents descriptive statistics of the investigated response-variables of materials made of resin and sawdust.

Results and Discussion

TABLE 4.1: Physical and mechanical properties of composite resin-sawdust.

CE	Statistics	MOE (MPa)	MOR (MPa)	ρ_{ap} (g/cm³)	A_m24h (%)
C1	X_m	2112	58.50	0.87	1.96
	DP	296.55	3.54	0.17	0.39
	CV (%)	14	6.04	19.90	20.00
C2	X_m	1104	36	0.94	4.01
	DP	140.22	2.83	0.16	0.84
	CV (%)	12.70	7.85	17.44	20.95
C3	X_m	2540	62	0.98	3.49
	DP	254.56	11.31	0.01	0.31
	CV (%)	10.02	18.24	1.73	8.93
C4	X_m	968	30	0.98	5.66
	DP	124.74	2.12	0.14	1.07
	CV (%)	12.88	7.19	14.83	18.90
C5	X_m	1,811	44.5	1.04	2.68
	DP	142.84	3.53	0.082	0.43
	CV (%)	7.89	7.94	7.88	16.11
C6	X_m	906	25	1.01	3.52
	DP	65.05	4.24	0.09	0.54
	CV (%)	7.18	16.97	9.07	15.59
C7	X_m	1.757	44	1.03	3.35
	DP	191.82	1.41	0.05	0.65
	CV (%)	10.91	3.21	5.82	19.49
C8	X_m	1,897	43	0.97	4.70
	DP	325.76	4.94	0.04	0.65
	CV (%)	17.33	11.64	4.61	13.99

Results and Discussion

According Sales *et al.* (2011), the compressive strength value of the compounds manufactured with cement matrix and waste from the water treatment and Pine wood were on average equal to 1.20 MPa, being 94.26% less than conventional concrete strength (20.9 MPa), classifying these materials, as enable for structural repairs. By the average values found of compressive strengths of manufactured resin-sawdust compounds [25;62 MPa],it follows that, according to Sales *et al.* (2011), the materials can also be classified as enable for structural repairs or even as reinforcement issues, to be the largest amount of mechanical strength (C3) 199.66% higher than conventional concrete resistance.

Table 4.2 presents the ANOVA results for the investigated physical and mechanical properties, finding underlined the smallest P-values than 0.05, considered significant at the 95% confidence level, being TS sawdust type, FG particle size range, FS sawdust fraction and TS*FG, TS*FS, FG*FS e TS*FG*FS the interactions between factors.

TABLE 4.2: ANOVA results for compound resin-sawing

	P-value							
Answers	**TS**	**FG**	**FS**	**TS*FG**	**TS*FS**	**FG*FS**	**TS*FG*FS**	**R^2 (Adj)**
MOE	0.380	0.012	0.000	0.127	0.001	0.242	0.003	89.47%
MOR	0.019	0.209	0.000	0.087	0.011	0.457	0.026	84.43%
ρ_{ap}	0.213	0.655	0.749	0.411	0.570	0.788	0.730	74.85%
A_m 24h	0.612	0.016	0.005	0.443	0.254	0.712	0.826	63.61%

Figure 4.1 presents the ANOVA waste charts in respect of physical and mechanical properties of the composite resin and sawdust, confirming normality in the distribution of waste by the P-values found are both greater than 0.05, helping to validate the model ANOVA.

Results and Discussion

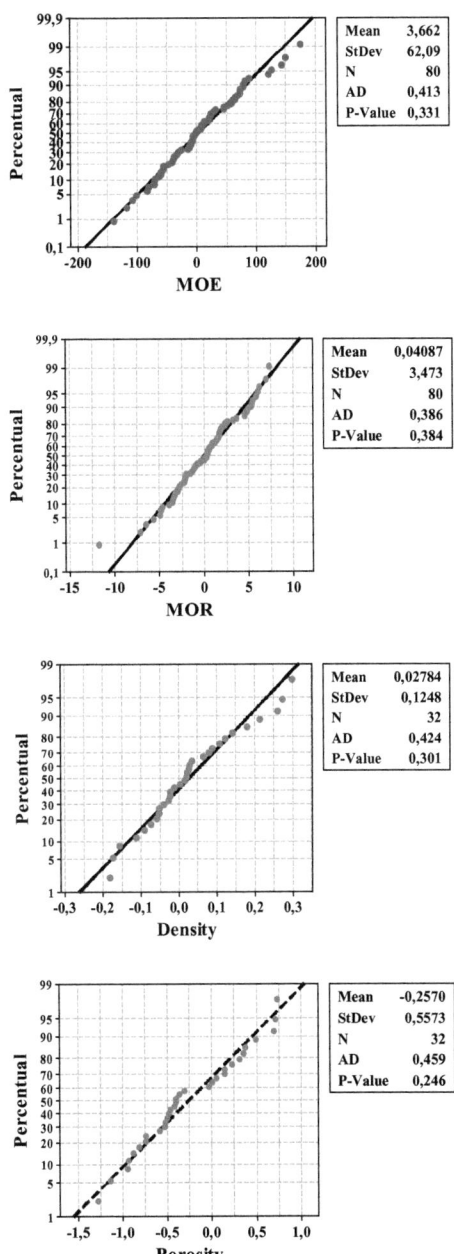

FIGURE 4.1: ANOVA waste chart about the physical and mechanical properties of composite resin-sawdust.

From Table 4.2, it is noticed that the interactions between the three factors investigated for MOE and MOR properties of the compounds resin-sawdust were significant. Figures 4.2 and 4.3 illustrate charts interactions of the major factors of the modulus of elasticity and compressive strength respectively.

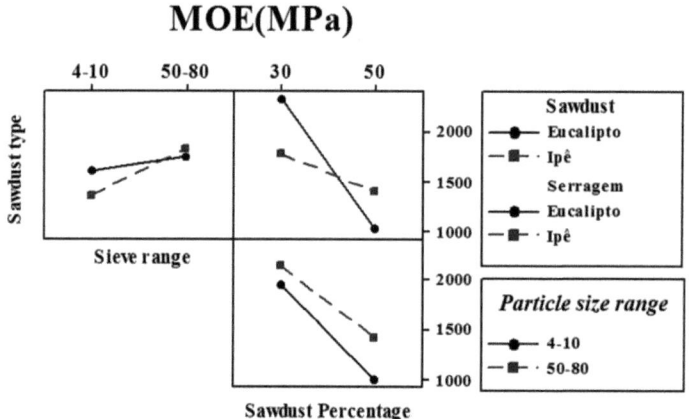

FIGURE 4.2: Chart of main effects on the MOE of resin-sawdust compounds.

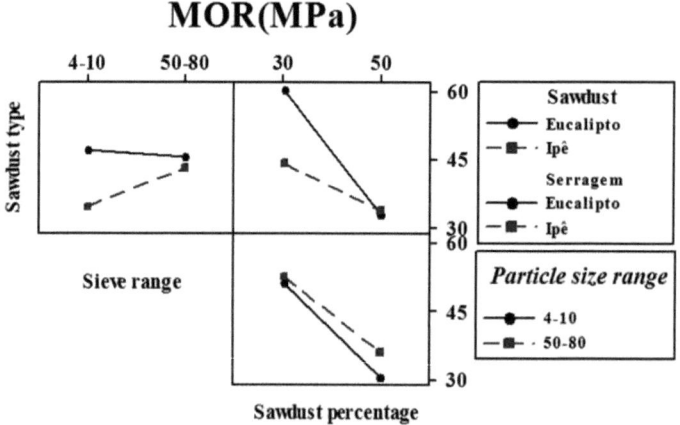

FIGURE 4.3: Chart of main effects on the MOR of the resin-sawdust compounds.

Results and Discussion

In general, for the MOE and MOR, the results of Figures 4.2 and 4.3 showed that the materials made with 30% of *Eucalyptus* wood sawdust in US-Tyler particle size range 50-80 presented the best results, being the C3 experimental condition to be used in the preparation of materials with the addition of Portland cement CP-V AIRI and magnesium silicate.

The ANOVA P-values on the apparent density showed that the individual factors and interactions were not significant, leading to statistically equivalent results. With respect to porosity, only individual factors: particle size range and fraction of sawdust were significant. Figure 4.4 illustrates the charts of the main factors for porosity.

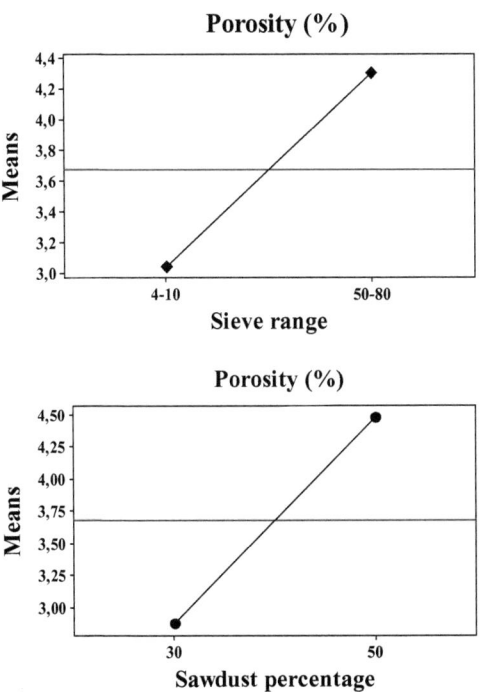

FIGURE 4.4: Chart of main effects on porosity (η)

Results and Discussion

Figures 4.4-a and 4.4-b show that the4-10 mesh and the fraction of 30% sawdust provided the best porosity values. Materials produced with use of mesh 4-10 showed to be 28% lower than elaborate mesh 50-80, and made with 30% of sawdust were 38% lower on average to those manufactured with 50% sawdust.

The Tabela4.3 presents descriptive statistics of the compound results of condition C3 with additions of cement and magnesium silicate.

TABLE 4.3: Results of physical and mechanical properties of the compounds with the addition of cement and talc

CE	Statistics	MOE (MPa)	ρ_{ap} (g/cm^3)	η (%)
	X_m	2,616	1.08	4.22
Cement (10%)	*DP*	253.72	0.26	0.65
	CV (%)	9.70	24.41	15.40
	X_m	2,778	1.12	4.96
Cement (20%)	*DP*	381.56	0.18	1.32
	CV (%)	13.74	16.07	26.61
	X_m	2,370	1.11	3.78
Talc (10%)	*DP*	296.43	0.20	0.79
	CV (%)	12.51	18.01	20.90
	X_m	2,463	1.13	4.08
Talc (20%)	*DP*	336.08	0.11	0.93
	CV (%)	13.64	9.73	22.79

The ANOVA P-value of the cement fraction factor on the stiffness of the manufactured material was equal to 0.029 [R^2 (Adj) = 83%], implying be significant the use of the cement particles in the MOE compounds. Figure 4.5 illustrates the waste graphic, helping to validate the model ANOVA.

Results and Discussion

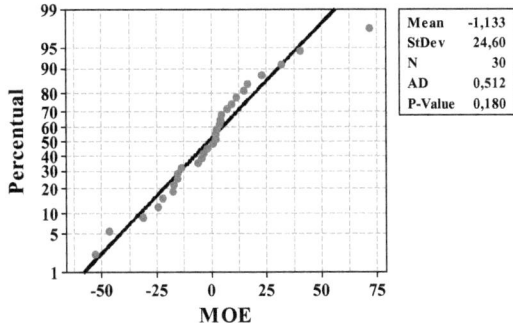

FIGURE 4.5: ANOVA of cement fraction factor on the MOE.

Figure 4.6 shows the chart of main effects of factor cement fractions of the elasticity modulus in compression of the resin-sawdust cement compounds. The inclusion of 10% by mass cement fraction afforded 4.15% increase relative to the reference condition (C3), and 9.70% higher than the stiffness of the material made of 20% cement in relation to the condition C3.

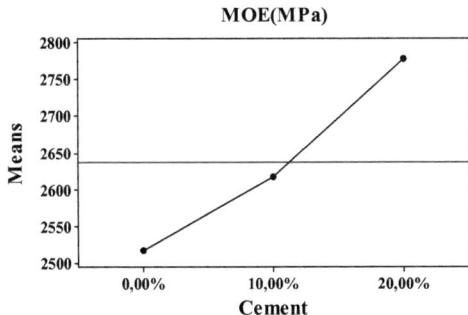

FIGURE 4.6: Chart of main effects of cement fraction factor on the MOE.

Table 4.4 presents the results of the Tukey Test cement fraction factor on the stiffness of the material made from the reference condition.

Results and Discussion

TABLE 4.4: Tukey Test for the MOE of materials with cement addition.

	MOE$_{0\%}$	MOE$_{10\%}$	MOE$_{20\%}$
Average (MPa)	2,540	2,616	2,778
Grouping	B	B	A

From Table 4.4, the grouping revealed that the materials made with and without 10% of cement results equivalent to the elasticity modulus in compression, being not equivalent or higher than MOE of the materials made with 20% in relation to cement, with 10 % and 0% of this reinforcement.

The ANOVA P-value for the apparent density of the materials produced with the addition of cement was equal to 0.000 [R^2 (Adj) = 76%], being the inclusion of cement significant in the density of fabricated materials. Figure 4.7 shows the waste normality chart.

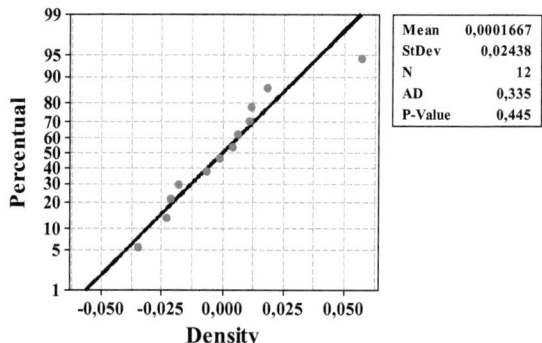

FIGURE 4.7: Chart of ANOVA waste of the apparent density of the materials produced with the addition of cement.

Figure 4.8 shows chart of the main effects off actor cement fractions of apparent density of compounds. The inclusion of 10% in mass cement fraction resulted in an increase

Results and Discussion

of 13.54% compared to reference condition (C3), being greater 17.70% than ρ_{ap} of made

materials of 20% cement in relation to the condition C3.

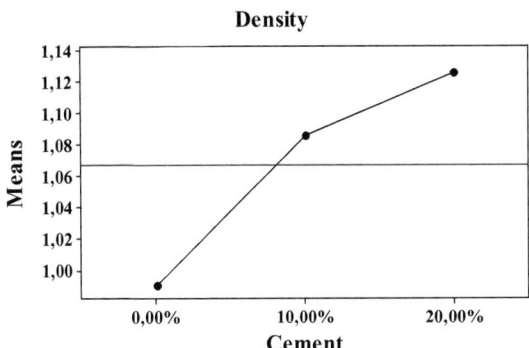

FIGURE 4.8: Chart of main effects of cement fraction factor on apparent density of the

materials manufactured with cement.

Table 4.5 presents the results of the Tukey Test of factor cement fractions on the

density of the materials manufactured of reference condition. The results indicated

equivalence between the use of fractions 10 and 20% of cement in the compounds,

presenting be not equivalents (above) to the reference condition (smallest average value).

TABLE 4.5: Tukey Test for the apparent density of materials with cement addition.

	ρ_{ap} 0%	ρ_{ap} 10%	ρ_{ap} 20%
Average (g/cm³)	0.98	1.08	1.12
Grouping	B	A	A

The ANOVA P-value for the porosity of the material manufactured with the addition

of cement was equal to 0.000 [R^2(Adj) = 66%], being the inclusion of this significantly

reinforcement in the porosity of the manufactured materials. Figure 4.9 shows the waste

normality chart.

Results and Discussion

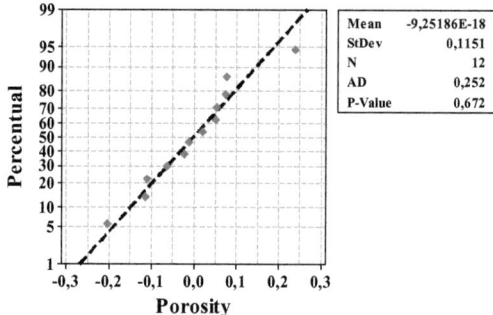

FIGURE 4.9: Chart of ANOVA waste on the porosity of the material manufactured with the addition of cement.

Figure 4.10 illustrates the chart of main effects of factor cement fractions on the porosity of the compounds. The inclusion of 10% in mass cement fraction gave a 14% increase relative to the reference condition (C3), and 36.62% greater than η materials made with 20% cement in relation to the condition C3.

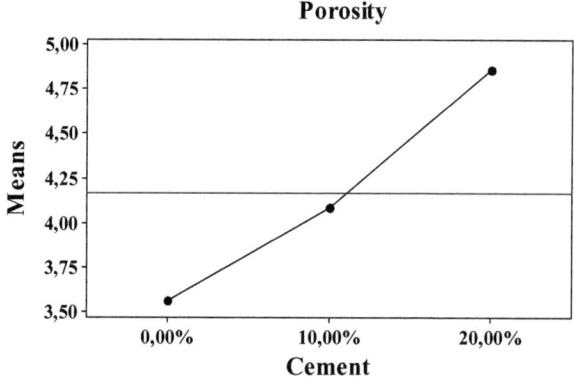

FIGURE 4.10: Chart of main effects of factor cement fraction on the porosity of the materials made of cement.

Results and Discussion

Table 4.6 presents the results of Tukey Test of factor cement fraction on porosity of the manufactured materials. The results indicated that the materials showed the highest porosity values were made with 20% of cement, followed by materials fabricated with the use of 10% to 0% of such reinforcement.

TABLE 4.6: Tukey Test for η of materials with cement addition.

	η 0%	η 10%	η 20%
Average (%)	3,49	4,22	4,96
Grouping	C	B	A

The ANOVA P-value of magnesium silicate fraction factor on the stiffness of the materials made on the reference condition C3 was equal to 0.371 [R^2 (Adj) = 81%], implying not mean the employment of talc in percentages MOE of the compounds. The Figure 4.11 shows the waste chart obtained from ANOVA.

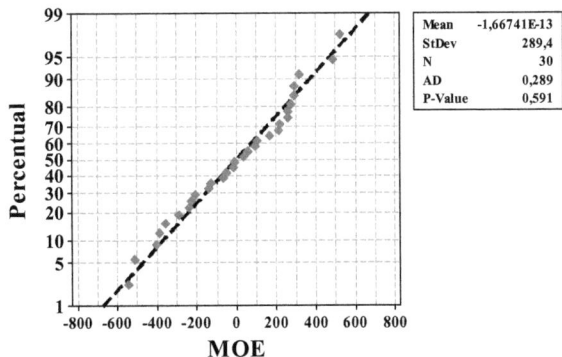

FIGURE 4.11: Chart of ANOVA waste of the MOE fabricated from materials manufactured with talc addition.

Results and Discussion

    The ANOVA P-value of magnesium silicate fraction factor on apparent density of the material manufactured, was equal to 0.192 [R^2 (Adj) = 72%], implying not be significant the use of silicate particles in density of compounds. The Figure 4.12 shows the chart of obtained residue ANOVA.

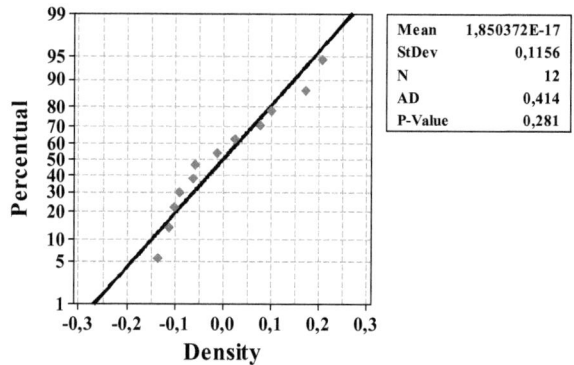

FIGURE 4.12: Chat of ANOVA waste of apparent density of the materials produced with talc addition

    The ANOVA P-value of magnesium silicate fraction factor on apparent density of the manufactured materials was equal to 0.002 [R^2 (Adj) = 63%], implying be significant use of talc particles in the porosity of the compounds. The Figure 4.13 shows the waste chart obtained from ANOVA.

Results and Discussion

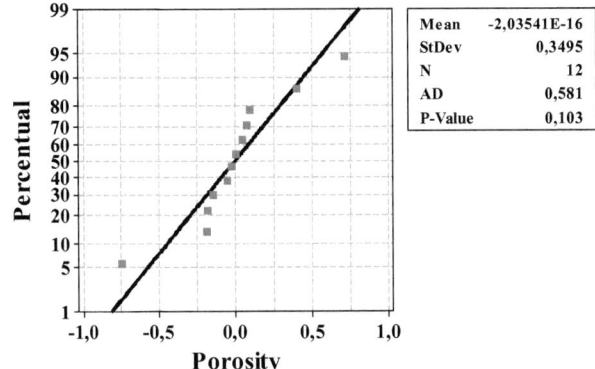

FIGURE 4.13: Chart of ANOVA waste of porosity of the material manufactured with the talc addition.

The Figure 4.14 represents the chart of main effects of factor talc fractions on the porosity of the compounds. The inclusion of 10% by mass fraction of magnesium silicate provided an increase of 10.73% compared to reference condition (C3), and 18.64% greater than η materials made with 20% of talc with respect to the C3 condition.

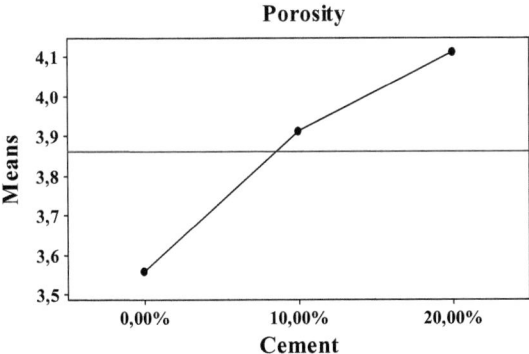

FIGURE 4.14: Chart of main effects of talc fraction factor on porosity.

Results and Discussion

Table 4.7 presents the results of Tukey Test of talc fraction factor on porosity of manufactured materials from the reference condition. The results indicated that the materials 10 and 20% talc particles were statistically equivalent, being higher than the porosity of the reference material condition (low value).

TABLE 4.7: Tukey Test for η materials with silicate addition.

	η 0%	η 10%	η 20%
Average (%)	3.49	3.78	4.08
Grouping	B	A	A

Chapter 5

CONCLUSIONS

The manufacturing process of the compounds developed, proved to be simple, being the materials obtained from direct mixing between the phases. For improving the physical and mechanical properties investigated, it was used a piece with usefulness that provide light compression, which can surely be reproduced with other equipment in the case of the production and use of this material for specific purposes.

The progressive incorporation of sawdust, as expected, gave reductions in the elasticity modulus and strength to compression of the composite.

The type of sawdust (wood species) was significant in the mechanical properties of composites. The species of *Eucalyptus*, whose elasticity modulus is lower than the Ipê wood, presented the highest values of mechanical properties, possibly justified by its better affinity with epoxy resin.

The particle range sizes and sawdust fractions were significant in porosity of manufactured resin-sawdust materials, presenting the compounds with finer mesh particles and the highest mass fraction of sawdust the highest values.

From the manufactured resin-sawdust composites, the treatment with 30% of *Eucalyptus* sawdust and 50-80 US-Tyler particle size range, showed the best results for the elasticity modulus in compression, which is the base condition (reference) chosen for inclusion of cement particles and talc.

The inclusion of cement mass fractions is shown to be significant in all the variables-response investigated, giving an increase in elasticity modulus in compression

Conclusions

The inclusion of magnesium silicate particles in the compounds of the base condition was only significant in porosity, not being in elasticity modulus, neither in apparent density. Thus, the magnesium silicate inclusion in the investigated fractions did not appear as good solution, as well as, presenting results equivalent to the resin-sawing compounds, so had value greater than the sawdust.

The suggestion is to compare the results with those of two other authors and classify the form of use of materials developed here.

In this line of thought, for future studies, it is expected to perform physical or chemical treatment in the wooden grains, in order to improve the interface conditions, to study new proportions with the addition of cement in these materials as well as the use fiber assemblies and distinct resins.

REFERENCES

Ahmad, I.; Mosadeghzad, R.; Daik, A.; Ramil, S. The effect of alkali treatment and filler size of the properties of sawdust/UPR composites based on recycled PET wastes. Journal of Applied Polymer Science, v.109, n.6, p. 3651-3658, 2009.

Ajiwe, V. I. E., Okeke, C. A.; Ekwuozor, S. C. A pilot plant for production of ceiling board from rice husks. Biores Technology, n.66, p.41-43, 1998.

Akgüla, M.; Çamlibelb, O. Manufacture of medium density fiberboard (MDF) panels from rhododendron (*R. ponticum L.*) biomass. Building and Environment. Part Special: Building Performance Simulation, Volume 43, Issue 4, pp. 438–443, 2008.

Al-Qurerish, H. A.; Silva, J. L. G. Mechanical soft wetting system of natural fiber with polymeric resin. Journal Processing Technology, v.92, n.93, p. 124-128, 1999.

American Society for Testing and Materials - ASTM D1921. Standard Tests Methods for Particle Size (Sieve Analysis) of Rigid Plastics, 2012.

American Society for Testing and Materials - ASTM D695-10:2010 Standard Test Methods for Compressive Properties of Rigid Plastics.

Ashori, A.; Nourbakhsh, A. Characteristics of wood-fiber plastic composites made of recycled materials, Waste Management, v.29 (4). p.1291-1295, 2009.

Bledzik, A. K.; Letman, M.; Viksne, A.; Rence, L. A comparison of compounding process and wood type for wood fibre-pp composites, Composites: Part A – Applied Science and Manufacturing, v.36, p.789-797, 2005

Bledzki A. K.; Gassan, J. Composites reinforced with cellulose based fibres. Progress and Polymer Science, v.24, p.221-274, 1999.

Brandt, C. W.; Fridley, K. J. Effect of load rate on flexural properties of wood-plastic composites. Wood and fiber Science, v.35, n.1, p.135-147, 2003.

References

Brazilian Association of Technical Standard - ABNT NBR 7190:1997. Design of timber structures. Rio de Janeiro, 1997.

Brazilian Association of Technical Standard - ABNT NBR 14810. Particleboards: 1, 2, 3. 2002.

Carvalho, L. H. Polymeric Composites Reinforced by Vegetable fibers. UFCG. 2003. Available at: www.abpol.com.br/apostila, accessed in March 2012.

Clemons, C. M. Interfacing Wood-plastic composites industries in the U.S.A. Forest Products Journal. 2002 – Available in site: www.jobwerx.com/news/Archives/iwpc.html, - accessed in March 2004.

Correa, C. A.; Fonseca, C. N. P.; Neves, S.; Razzino, C. A.; Hage, E. Jr. Thermoplastic composites with wood, Polymers: Science and Technology, Department of Materials Engineering, São Carlos Federal University, São Carlos/SP. Vol. 12, n. 3, p. 154-165, 2003.

Dagher, H. J.; Bragdon, M. M.; Lindyberg, R. F. Advanced fiber reinforcement polymer wood composites in transportation applications. Transportation Research Board National Research Council. University of Maine, Orano - USA. nº. 1814, pp. 237-242, 2002.

Dias, F. M. Application of polyurethane resin on mammon base in the manufacture of particleboard panels. Book Chapter: LAHR, F.A.R. Wooden derived products. São Carlos: EESC/USP, p. 37-160, 2008.

Dibenedetto, A. T. Tailoring of interfaces in glass fiber reinforced polymer composite: a review, Materials Science and Engineering, v.32 (1), p.74-82, 2001.

English, B.; Clemons, C. M. Weight reduction: wood versus fillers in polypropylene. Proceeding of the fourth international conference on wood fiber, plastic composites, 237-244, Madison, Wisconsin - USA, 1997.

European Committee for Standardization - EN 317. Particleboards and fibreboards - Determination of swelling in thickness after immersion in water. Brussels, 1993.

European Committee for Standardization - EN 323. Wood-based panels - Determination of density. Brussels, 1993.

References

Garcez, M. R.; Meneghetti, L. C.; Cauduro, L. B.; Campagnolo, J. L.; Silva, F. LCP reinforcement structures with fiber-based polymers, 2nd Pathology Seminar of Buildings. New materials and emerging technologies, Rio Grande do Sul Federal University, Porto Alegre/RS. Annals of Pathology II Seminar of Buildings, From 18 to 19 November 2004.

Hisham, S.; Faieza, A. A.; Ismail, N; Sapuan, S. M.; Ibrahim, M. S. Flexural mechanical characteristic of sawdust and chipwood filled epoxy composites – Key Engineering Materials, Composite Science and Technology, Department of Mechanical and Manufacturing Engineering University Putra Malasya, Serdang, Selangor, Malasya, , Vol. 471-472, p 1064-1069, 2011.

Hisham, S.; Faieza, A. A.; Ismail, N; Sapuan, S. M.; Ibrahim, M. S. Tensile properties and micro morphologies of sawdust and chipwood filled epoxy composites. Key Engineering Materials, Composite Science and Technology, University Putra Malasya, Serdang, Selangor, Malasya. Vol. 471-472, pp. 1070-1074, 2011.

Hristov, V.N.; Krumova, M.; Vasileva, S.; Michler, G. H. – Modified polypropylene wood flour composites. II Fracture, deformation and mechanical properties. Journal Applied Polymer Science, v.9, n. 2, p. 1286-1292, 2004.

Jayaraman, K.; Bhattacharyya, D. Effects of plasma treatment in enhancing the performance of wood fibre-polypropylene composites, Composites Part A: Applied Science and Manufacturing, Volume 35, Issue 12, December 2004, Pages 1363-1374 – 2004.

Joseph, K.; Medeiros, E. S.; Carvalho, L. H. Composites of polyester matrix reinforced with short sisal fibers - Polymers: Science and Technology, v.9, n.4, p. 136-140 -1999.

Kamel, S.; Abeer, M. A.; Mohamed, El-S.; Zenat, A. N. Mechanical properties and water absorption of low density polyethylene/sawdust composites, Journal of Applied Polymer Science, v.107(2), p. 1337-1342, 2008.

Khoo T. S.; Ratnam, M. M.; Shahnaz, S. A. B.; Abdul Khalil, H. P. S. I. Wood Filler-recycled Polypropylene (WF-RPP) Composite Pallet: Study of Fastening Method, Journal of Reinforced plastics and Composites, Vol. 27, No. 16–17, 2008.

References

Khoylou, F. Radiation-induced polymer impregnated sawdust-clay-cement composite – Polymers and Polymer Composite, v.14, n.8, p. 826-831, Rapra Technology Ltd., 2006.

Koenig, K. M.; Sypkens, C.W. Wood-plastic composites for market shore – Wood and Wood Products, v.107, n.5, p.49-58, 2002.

Ku, H., Prajapati, M.; Cardona, F. Thermal properties of sawdust reinforced vinyl ester composites post-cured in microwaves: pilot study, Composites Part B, 2011, Vol. 42, pp. 898-906, 2011.

Kuruvilla, J.; Tolêdo, R.D.; Beena, J.; Sabu, T.; Carvalho, L. H. A review on sisal fiber reinforced polymer composites, Journal of Agricultural and Environmental Engineering, v.3, n.3, p.367-379, 1999.

Mchenry, E.; Stachurski, Z. H. Composite Materials Based on Wood and Nylon Fibre, Composites Part A: Applied Science and Manufacturing, 34(2): 171–181, 2003.

Mioto, J. L .; Dias, A. A. Reinforcement and recovery of timber structures, Seminar Exact and Technological Sciences, Londrina/PR. Vol. 27, No 2, pp. 163-174, 2005.

Missagia, Z. M. V.; Santos, J. C.; Christoforo, A. L.; Panzera, T. H.; Silva, V. R. V. Compressive Behaviour of Polimeric Composites Reinforced with Sawdust. In: Brazilian Conference on Composite Material (BCCM), 2012, Belo Horizonte. Annals of the First Brazilian Conference on Composite Material, 2012.

Nassar, M. A. Obtaining some polymer composites filled with rice husks ash, Polymer Plastics technology and Engineering, v.46, n.5, p. 441-446, 2007.

Paes, J. B.; Nunes, S. T; Rocco Lahr, F. A.; Nascimento, M. F.; Lacerda, R. M. A. Quality of sheets of glued *Pinus elliottii* particles with polyurethane resin under different combinations of pressure and temperature. Forest Science, Santa Maria/RS, v. 21, n. 3, p. 551-558, 2011.

Panthapulakkal, S.; Law, S.; Sain, M. Enhancement of processability of rice-husk filled high-density polyethylene profiles – Journal of Thermoplastic composite Materials, v.18, n.5, p. 445-458, 2005.

References

Passerotti, G. F. A.; Campos, C. I.; Nascimento, M. F.; Rocco Lahr, F. A. Characterization and production of particulate panel of eucalyptus sp, using polyurethane adhesive. In: 18th Brazilian Congress of Engineering and Material Sciences, CBECiMat. Porto de Galinhas/ PE from 24 to 28 November 2008.

Rangel, G. W. A.; Souza Jr, D. A.; Gesualdo, F. A. R.; Santos, A. C.; Barreiro, C. H. Numerical Analysis - Experimental of Timber Reinforced Beams by PRFC. ENTAC. Thirteenth National Meeting of Built Environmental Technology. Canelas/RS- RS. From 6 to 8 October 2010.

Saffian, H. A.; Harun, J.; Thair, P. M; Abdar, K. Feasibility of Manufacturing a Medium Density Fibreboard Made of 4-Year Old Rubber Tree RRIM 2020 Clone. Key Engineering Materials - Composite Science and Technology, Vol. 471, pp. 839-844, 2011.

Salemane, M. G.; Luyt, A. S. Thermal and mechanical properties of polypropylene-wood powder composite – Journal Appl. Polym Science, n.100, p.4173-4180, 2006.

Sales, A.; Souza, F. R.; Almeida, F. C. R. Mechanical properties of concrete produced with a composite of water treatment sludge and sawdust, Construction and Building Materials, 25(2) 2793–2798, 2009.

Salvastano, H. Natural fibers for producing construction components. International Composite of Fib reinforced Materials, Cali, Colombia, Del Valle University. Cyted Poyecto VIII. 5, 1998.

Sanches, E. Notes on structural reinforcement with carbon fiber sheet. Engineering, study and research. Juiz de Fora Federal University, pp. 67-73, Juiz de Fora/MG, 2001.

Savastano, H.; Agopyan, V.; Senff, L., Senff, L. Characterization of cement composite with addition of wood particles - UDESC - Santa Catarina State University, - Master's thesis, 2004.

Silva, S.; Rocco Lahr, F.A. Particles sheets made with waste of tropical low-density woods. Book: Waste recycling for civil construction. FUMEC University Publisher. Chapter 14, p. 343-365. Belo Horizonte/MG, 2007.

References

Stark, N. Influence of moisture absorption on mechanical properties of wood flour-polypropylene composites. Journal of Thermoplastic Composite Materials, v.14, n.5, p.421-432, 2001.

Tang, L.; Qu, B.; Shen, Y. Mechanical properties, morphological structure and thermal behavior of dynamically photo cross linked PP/EPPM blends. v.92, p.3371-3380, 2004.

Teixeira, M. G. Application of concepts of industrial ecology for the production of ecological materials. The example of the wood waste. Master's thesis, Polytechnic School, Bahia Federal University, Salvador/BA, 2005.

Vijay, P.V.; Gangarao, H. V. S.; Liang, R.; Skidmore, M. Rapid restoration of rail road timber bridges using polymer composites. Annual Technical Conference - ANTEC, Conference Proceedings. 69th Annual Technical Conference of the Society of Plastics Engineers, ANTEC, 2011.

Werkema, M. C. C.; Aguiar, S. Planning and Analysis of Tests: how to identify and assess the main influential variables in a process. Christiano Ottoni Foundation, UFMG Engineering School, Belo Horizonte/MG, 2005.

Yamaji, F. M.; Boudelle, A. Use of sawdust in the production of wooden composite plastic. Forest Journal, Curitiba/PR. Vol. 39, n.1, pp. 54-66, 2004.

Yuan, X. W.; Jayaraman, K.; Bhattacharyya, D. Effects of plasma treatment in enhancing the performance of wood-fiber polypropylene composites, Composites Parte A: Applied Science and Manufacturing, v.35 (12), p. 1363-1374, 2004.

Printed by Books on Demand GmbH, Norderstedt / Germany